The Intelligent Future: Exploring the Frontiers of Artificial Intelligence

Welcome to "The Intelligent Future: Exploring the Frontiers of Artificial Intelligence." In this book, we embark on an exciting journey into the world of artificial intelligence (AI) and its transformative potential. AI has emerged as a groundbreaking technology that is reshaping various aspects of our lives, from industries and economies to healthcare and everyday experiences.

In this rapidly evolving field, we delve into the latest advancements, cutting-edge research, and real-world applications of AI. We will explore the fundamental concepts, underlying principles, and ethical considerations surrounding AI. Through engaging storytelling and insightful discussions, we aim to demystify AI and empower readers to understand its impact on society.

Whether you are a curious individual seeking to expand your knowledge or a professional aiming to navigate the AI landscape, this book serves as your guide. Together, we will uncover the possibilities and limitations of AI, delve into its various subfields such as machine learning and natural language processing, and examine how AI is revolutionizing industries like healthcare, finance, and transportation.

Through practical examples, case studies, and expert insights, we will address the challenges and opportunities that arise with the rise of AI. We will also discuss the ethical considerations and responsible practices necessary to ensure AI is developed and deployed for the benefit of humanity.

As we embark on this journey, let us embrace the potential of AI and its ability to shape the future. Join us as we explore the frontiers of artificial intelligence and discover the remarkable possibilities that lie ahead.

Are you ready to dive into "The Intelligent Future"? Let's begin our exploration of this remarkable field together.

I. Introduction

The Transformative Power Of Artificial Intelligence

Artificial Intelligence (AI) has emerged as a transformative force, revolutionizing various aspects of our lives. It has the power to reshape industries, redefine human capabilities, and unlock new possibilities. The transformative power of AI lies in its ability to process vast amounts of data, learn from patterns, and make intelligent decisions, all with incredible speed and accuracy.

One of the most profound impacts of AI is in the realm of automation. AI-powered systems and algorithms can perform tasks that traditionally required human intervention, leading to increased efficiency, productivity, and cost savings across industries. From manufacturing and logistics to customer service and healthcare, AI is streamlining processes and driving innovation.

AI also has the potential to enhance decision-making and problem-solving. By analyzing complex data sets and providing valuable insights, AI enables us to make informed choices and predictions. It empowers professionals in fields such as finance, marketing, and research to uncover hidden patterns, identify trends, and make strategic decisions with greater precision.

Furthermore, AI is revolutionizing personalized experiences. Through machine learning algorithms and data analytics, AI systems can understand individual preferences, behaviors, and needs, enabling businesses to deliver personalized recommendations, tailored products, and customized services. This level of personalization enhances customer satisfaction and strengthens brand loyalty.

In the healthcare industry, AI is transforming diagnostics, drug discovery, and patient care. AI algorithms can analyze medical images, detect anomalies, and aid in early disease detection. Virtual assistants powered by AI can provide personalized health advice and support, improving patient outcomes and empowering individuals to take control of their well-being.

AI is also driving advancements in fields like transportation, robotics, and space exploration. Autonomous vehicles are becoming a reality, robotic assistants are improving efficiency and safety in various industries, and AI is helping us understand the complexities of our universe.

However, as we embrace the transformative power of AI, we must also consider its ethical implications. Issues like data privacy, algorithmic bias, and job displacement require careful consideration and responsible development of AI systems. It is crucial to ensure transparency, fairness, and accountability in AI applications, while prioritizing human well-being and societal benefits.

The transformative power of AI is vast and continues to expand as technology evolves. It holds the potential to revolutionize industries, enhance human capabilities, and solve complex challenges. As we harness this power, it is essential to approach AI with a balanced perspective, leveraging its capabilities for the greater good while addressing its ethical, social, and economic implications. By doing so, we can truly unlock the transformative potential of AI and shape a future that benefits all of humanity.

Overview Of The Book's Purpose And Scope

"The Intelligent Future: Exploring the Frontiers of Artificial Intelligence" is a comprehensive guide that aims to provide readers with a deep understanding of artificial intelligence (AI) and its transformative potential. The book covers a wide range of topics related to AI, including its foundations, real-world applications, ethical considerations, advancements, and societal impact.

The purpose of this book is to demystify the field of artificial intelligence and showcase its significance in shaping our future. By exploring the frontiers of AI, readers will gain insights into its underlying technologies, such as machine learning, deep learning, and natural language processing. They will also discover how AI is being applied in various industries, including healthcare, finance, transportation, retail, and manufacturing.

Ethical considerations in AI are given due importance, with discussions on bias, privacy, transparency, and responsible practices. The book also delves into the advancements and future trends in AI, such as reinforcement learning, generative AI, robotics, and the intersection of AI with the Internet of Things (IoT) and quantum computing.

Furthermore, the book addresses the impact of AI on society, covering topics like job displacement, education, ethics, policy, and the potential for AI to contribute to social good and address global challenges.

Through a balanced blend of technical explanations, real-world examples, and thought-provoking discussions, this book aims to empower readers to understand the potential of AI and its implications. Whether you are a student, professional, or simply curious about AI, "The Intelligent Future" offers a comprehensive exploration of AI's frontiers and inspires readers to embrace the opportunities and challenges of an intelligent future.

II. Understanding Artificial Intelligence

Defining Artificial Intelligence

Artificial intelligence (AI) refers to the development and deployment of computer systems and algorithms that can perform tasks that typically require human intelligence. It involves the creation of intelligent machines that can perceive, reason, learn, and interact with their environment to accomplish specific goals. AI encompasses a broad range of technologies, including machine learning, natural language processing, computer vision, robotics, and more.

At its core, AI aims to replicate and augment human cognitive abilities, enabling machines to process and analyze vast amounts of data, recognize patterns, make decisions, and even engage in complex problem-solving. It allows computers to simulate human intelligence, adapt to changing circumstances, and continually improve their performance through learning from experience.

AI has the potential to revolutionize various industries and domains, from healthcare and finance to transportation and entertainment. It enables the automation of repetitive tasks, the enhancement of decision-making processes, the development of personalized user experiences, and the advancement of scientific research. By leveraging AI, businesses and organizations can drive innovation, improve efficiency, and unlock new possibilities for solving complex problems.

Historical Context And Evolution Of Ai

The field of artificial intelligence (AI) has a rich historical context and has undergone significant evolution over the years. The origins of AI can be traced back to the 1950s when researchers began exploring the possibility of creating machines that could mimic human intelligence. The term "artificial intelligence" was coined in 1956, marking the official birth of the field.

In its early years, AI focused on rule-based systems and symbolic processing, where computers followed explicit instructions to solve problems. However, progress was limited due to the complexity of real-world problems and the lack of computational power and data.

The field experienced a resurgence in the 1980s with the advent of machine learning, a subfield of AI that enables computers to learn from data and improve their performance over time. This approach shifted the focus from explicit programming to training algorithms on large datasets, allowing machines to discover patterns and make predictions.

In recent years, AI has seen remarkable advancements, thanks to breakthroughs in deep learning and neural networks. These techniques, inspired by the structure and function of the human brain, have revolutionized AI by enabling computers to process vast amounts of data, extract meaningful insights, and perform tasks with unprecedented accuracy. Today, AI is integrated into various applications, from voice assistants and autonomous vehicles to healthcare diagnostics and financial trading.

The evolution of AI has been driven by advancements in computing power, data availability, and algorithmic innovation. As technology continues to progress, AI holds immense potential to transform industries, drive innovation, and shape the future of human society.

Types Of Artificial Intelligence Systems

Artificial intelligence (AI) systems can be categorized into different types based on their capabilities and functionalities. Here are some common types of AI systems:

Reactive AI: Reactive AI systems are designed to react to specific situations without any memory or learning capabilities. They analyze the current state and respond based on predefined rules or patterns. These systems excel at tasks with clear objectives, such as playing chess or providing automated customer support.

Limited Memory AI: Limited Memory AI systems can store and recall past data to make informed decisions. They have the ability to learn from historical data and use that information to improve their future actions. These systems are commonly used in recommendation engines, fraud detection, and natural language processing applications.

Theory of Mind AI: Theory of Mind AI refers to systems that have the ability to understand and attribute mental states to others. They can infer the thoughts, intentions, and emotions of individuals and use that understanding to interact and communicate effectively. Although still in the early stages of development, these systems hold promise for applications such as social robotics and human-like chatbots.

Self-Aware AI: Self-Aware AI represents the highest level of AI sophistication, where machines possess consciousness and self-awareness. This level of AI is purely speculative and currently exists only in science fiction. It involves machines having a sense of their own existence, emotions, and consciousness.

It's important to note that the categorization of AI systems is not always rigid, and there can be overlaps between these types. Additionally, AI technologies continue to evolve, and new types of AI systems may emerge as research progresses.

III. *Foundations of Artificial Intelligence*

Machine Learning: Algorithms And Techniques

Machine learning is a subset of artificial intelligence that focuses on enabling computers to learn and improve from data without being explicitly programmed. It involves the use of algorithms and techniques that allow machines to automatically discover patterns, make predictions, and generate insights from large datasets. Here are some common algorithms and techniques used in machine learning:

Supervised Learning: Supervised learning algorithms learn from labeled data, where the input data is paired with corresponding output labels. The algorithms analyze the data to identify patterns and relationships, enabling them to make predictions or classify new, unseen data.

Unsupervised Learning: Unsupervised learning algorithms work with unlabeled data, where the algorithm seeks to discover inherent patterns and structures without any predefined labels. Clustering algorithms, such as k-means clustering and hierarchical clustering, are commonly used in unsupervised learning.

Reinforcement Learning: Reinforcement learning involves training an agent to make sequential decisions in an environment to maximize rewards. The agent learns through trial and error, receiving feedback in the form of rewards or penalties based on its actions. This technique is often used in robotics, game playing, and autonomous systems.

Deep Learning: Deep learning is a subfield of machine learning that focuses on training artificial neural networks with multiple layers to process complex patterns and large amounts of data. Deep learning has achieved remarkable success in various domains, including image and speech recognition, natural language processing, and autonomous vehicles.

Transfer Learning: Transfer learning allows models trained on one task or dataset to be applied to another related task or dataset. By leveraging knowledge learned from one domain, transfer learning can accelerate the training process and improve performance, particularly in scenarios with limited labeled data.

Ensemble Learning: Ensemble learning combines multiple models to improve the overall prediction or classification accuracy. Techniques such as bagging, boosting, and stacking are commonly used to create diverse models and aggregate their predictions.

These are just a few examples of the many algorithms and techniques available in machine learning. The choice of algorithm depends on the specific problem, the type and size of the dataset, and the desired outcome. Machine learning continues to advance, and new algorithms and techniques are constantly being developed to address complex challenges.

Deep Learning: Neural Networks And Their Applications

Deep learning is a subfield of machine learning that focuses on training artificial neural networks with multiple layers to process complex patterns and large amounts of data. Neural networks are computational models inspired by the structure and functioning of the human brain. They consist of interconnected nodes, called neurons, which process and transmit information.

In deep learning, neural networks with many hidden layers are used to learn hierarchical representations of data. Each layer learns increasingly complex features, allowing the network to extract intricate patterns and relationships. Deep learning has gained significant attention and achieved remarkable success in various applications, including:

Image Recognition: Deep learning has revolutionized image recognition tasks. Convolutional Neural Networks (CNNs) are widely used for tasks such as object detection, image classification, and facial recognition.

Natural Language Processing (NLP): Deep learning models have greatly improved the performance of NLP tasks, such as language translation, sentiment analysis, and text generation. Recurrent Neural Networks (RNNs) and Transformers are commonly used in NLP.

Speech Recognition: Deep learning has played a crucial role in advancing speech recognition systems, enabling accurate speech-to-text conversion and voice assistants like Siri and Alexa. Recurrent Neural Networks (RNNs) and Convolutional Neural Networks (CNNs) are used in speech recognition models.

Autonomous Vehicles: Deep learning is instrumental in developing self-driving cars. Deep neural networks process sensor data, such as images from cameras and lidar, to make real-time decisions for navigation, object detection, and path planning.

Healthcare: Deep learning has shown promise in medical image analysis, disease diagnosis, and drug discovery. Neural networks can analyze medical images, such as MRI scans, and assist in diagnosing diseases like cancer.

Gaming and Robotics: Deep learning techniques, such as Reinforcement Learning and Deep Q-Networks, have been applied to develop intelligent agents for playing complex games and controlling robotic systems.

Deep learning continues to advance rapidly, with ongoing research focusing on improving model architectures, training algorithms, and handling large-scale datasets. Its ability to learn complex representations from data makes it a powerful tool for solving intricate problems across various domains.

Natural Language Processing: Understanding And Generating Human Language

Natural Language Processing (NLP) is a branch of artificial intelligence that focuses on the interaction between computers and human language. It encompasses the ability of machines to understand, interpret, and generate human language in a way that is meaningful and contextually relevant.

NLP involves a wide range of techniques and algorithms that enable machines to process and analyze text and speech data. Some key tasks in NLP include:

Text Classification: NLP algorithms can classify text into predefined categories or labels. This is useful for tasks such as sentiment analysis, spam detection, and topic classification.

Named Entity Recognition (NER): NER is the process of identifying and extracting named entities such as names of people, organizations, locations, and dates from text. It helps in information extraction and knowledge graph construction.

Sentiment Analysis: NLP models can determine the sentiment or emotion expressed in a piece of text, whether it is positive, negative, or neutral. This is valuable for understanding public opinion, customer feedback analysis, and brand monitoring.

Machine Translation: NLP algorithms can automatically translate text from one language to another. This is evident in popular translation services like Google Translate, which use sophisticated machine learning models to achieve accurate translations.

Question Answering: NLP systems can comprehend questions posed in natural language and provide relevant answers by analyzing large amounts of text data. This is employed in virtual assistants like Siri and chatbots.

Text Generation: NLP models can generate human-like text based on given prompts or conditions. This is applied in various applications such as chatbots, language generation for creative writing, and summarization of long texts.

NLP techniques utilize machine learning algorithms, statistical models, and linguistic rules to analyze and process language data. The field is constantly evolving, with ongoing research in areas like contextual language understanding, language generation, and the integration of NLP with other AI technologies like computer vision and speech recognition.

IV. Real-World Applications of Artificial Intelligence

Healthcare: Ai In Diagnostics, Treatment, And Drug Discovery

Artificial intelligence (AI) has made significant advancements in the healthcare industry, revolutionizing diagnostics, treatment, and drug discovery. AI-powered systems and algorithms have the potential to enhance medical decision-making, improve patient outcomes, and accelerate the development of new treatments. Here are some key applications of AI in healthcare:

Diagnostics: AI can analyze medical images, such as X-rays, MRIs, and CT scans, to assist in the detection and diagnosis of diseases. Machine learning algorithms can learn from vast amounts of image data to identify patterns and anomalies that may indicate specific conditions or diseases.

Treatment Planning: AI can help physicians develop personalized treatment plans by analyzing patient data, including medical records, genetic information, and clinical guidelines. AI algorithms can assist in identifying optimal treatment options based on individual patient characteristics, improving treatment efficacy and reducing adverse effects.

Drug Discovery: AI is revolutionizing the process of drug discovery by enabling faster and more efficient identification of potential drug candidates. Machine learning algorithms can analyze vast amounts of data, including genomic data, chemical structures, and clinical trial results, to identify drug targets, predict drug efficacy, and accelerate the development of new therapies.

Patient Monitoring: AI-powered wearable devices and remote monitoring systems can continuously collect and analyze patient data, such as vital signs, sleep patterns, and activity levels. This real-time monitoring allows for early detection of changes in health status, enabling timely interventions and personalized care.

Virtual Assistants and Chatbots: AI-based virtual assistants and chatbots are being used to provide personalized healthcare information, answer patient queries, and assist in triage. These tools can improve patient engagement, provide 24/7 support, and reduce the burden on healthcare providers.

Predictive Analytics: AI algorithms can analyze large datasets to identify patterns and trends that may predict disease progression, readmission rates, and treatment outcomes. This enables healthcare providers to proactively intervene, optimize resource allocation, and improve patient care.

The integration of AI in healthcare is still in its early stages, but it holds immense promise for transforming the industry. Ethical considerations, data privacy, and regulatory frameworks are essential in ensuring the responsible and effective implementation of AI technologies in healthcare settings.

Finance: Ai In Fraud Detection, Risk Assessment, And Trading

Artificial intelligence (AI) has brought significant advancements to the field of finance, particularly in areas such as fraud detection, risk assessment, and trading. AI-powered systems and algorithms have the potential to enhance security, improve decision-making, and optimize trading strategies. Here are some key applications of AI in finance:

Fraud Detection: AI algorithms can analyze large volumes of financial data, including transaction records, customer behavior patterns, and historical fraud cases, to detect and prevent fraudulent activities. Machine learning algorithms can learn from patterns and anomalies to identify suspicious transactions in real-time, reducing financial losses and protecting customer assets.

Risk Assessment: AI can assist in assessing and managing financial risks by analyzing historical data, market trends, and economic indicators. Machine learning algorithms can identify patterns and correlations that may indicate potential risks, enabling more accurate risk assessment and portfolio management.

Trading and Investment Strategies: AI algorithms can analyze vast amounts of financial data, including market prices, news articles, and social media sentiment, to identify patterns and trends that may impact investment decisions. Machine learning techniques can help develop trading models and optimize investment strategies, improving portfolio performance and generating more informed trading decisions.

Customer Service and Personalized Recommendations: AI-powered chatbots and virtual assistants can provide personalized financial advice and assistance to customers. These tools can analyze customer data, financial goals, and risk profiles to offer tailored recommendations, answer queries, and provide real-time financial guidance.

Algorithmic Trading: AI-driven algorithms can execute trades automatically based on predefined rules and market conditions. These algorithms can analyze market data, identify trading opportunities, and execute trades at high speeds, increasing efficiency and reducing the impact of human bias.

Credit Scoring and Underwriting: AI can enhance the accuracy of credit scoring models by leveraging alternative data sources, such as social media profiles and transaction histories. Machine learning algorithms can analyze these data points to assess creditworthiness and make more informed lending decisions.

The integration of AI in finance has the potential to transform the industry, improving efficiency, reducing risks, and enhancing customer experiences. However, it is important to ensure transparency, accountability, and ethical considerations in the use of AI technologies in finance. Regulatory frameworks and data privacy safeguards are crucial in maintaining trust and protecting consumer interests.

Transportation: Ai In Autonomous Vehicles And Smart Traffic Management

Artificial intelligence (AI) is revolutionizing the transportation industry, particularly in the areas of autonomous vehicles and smart traffic management. AI-powered systems and algorithms are driving advancements in safety, efficiency, and sustainability. Here are some key applications of AI in transportation:

Autonomous Vehicles: AI plays a crucial role in enabling self-driving cars and other autonomous vehicles. AI algorithms process data from various sensors, such as cameras, lidar, and radar, to perceive and interpret the surrounding environment. Machine learning techniques help vehicles learn from real-world scenarios, improving their ability to make informed decisions and navigate complex traffic situations.

Predictive Maintenance: AI algorithms can analyze vast amounts of sensor data from vehicles to predict maintenance needs and detect potential issues before they lead to breakdowns. By monitoring factors like engine performance, tire pressure, and battery health, AI systems can identify patterns and anomalies that may indicate upcoming failures or maintenance requirements, enabling timely repairs and reducing downtime.

Traffic Management: AI-based traffic management systems can optimize traffic flow, reduce congestion, and enhance transportation efficiency. By analyzing real-time data from sensors, cameras, and connected vehicles, AI algorithms can identify traffic patterns, predict congestion, and dynamically adjust signal timings to improve traffic flow. This can lead to reduced travel times, lower fuel consumption, and improved air quality.

Intelligent Route Planning: AI-powered route planning algorithms can consider various factors, such as real-time traffic conditions, weather, and road closures, to recommend the most efficient and fastest routes. These algorithms can take into account historical traffic data and learn from user preferences to provide personalized route recommendations and improve overall navigation experiences.

Fleet Management: AI can optimize fleet operations by analyzing data on vehicle usage, driver behavior, and maintenance schedules. Machine learning algorithms can identify patterns and trends to optimize routes, reduce fuel consumption, and enhance vehicle utilization. AI can also assist in driver scheduling, load optimization, and logistics planning, leading to cost savings and improved efficiency.

Passenger Experience: AI-powered systems can enhance the passenger experience by providing personalized recommendations, real-time updates, and improved accessibility. Virtual assistants and voice recognition technologies can assist passengers in navigating transportation options, accessing relevant information, and providing seamless interactions throughout their journey.

The integration of AI in transportation holds great potential for enhancing safety, efficiency, and sustainability in the industry. However, it is crucial to address challenges related to regulatory frameworks, cybersecurity, and public acceptance to ensure the successful deployment and adoption of AI technologies in transportation.

Retail: Ai In Personalized Marketing And Customer Experience

Artificial intelligence (AI) is transforming the retail industry by revolutionizing personalized marketing and enhancing the overall customer experience. With AI-powered systems and algorithms, retailers can better understand customer preferences, predict their needs, and deliver tailored experiences. Here are some key applications of AI in retail:

Customer Segmentation: AI algorithms can analyze vast amounts of customer data, including purchase history, browsing behavior, and demographics, to segment customers into distinct groups based on their preferences and behaviors. This allows retailers to target specific customer segments with personalized marketing campaigns and product recommendations.

Personalized Recommendations: AI-powered recommendation systems use machine learning algorithms to analyze customer data and generate personalized product recommendations. By considering factors such as past purchases, browsing history, and similarities with other customers, retailers can offer relevant and targeted recommendations, increasing customer engagement and driving sales.

Virtual Shopping Assistants: AI-powered virtual shopping assistants, chatbots, and voice assistants provide personalized and interactive customer support. These assistants can answer customer inquiries, provide product recommendations, and guide customers through the purchasing process, enhancing the overall shopping experience and improving customer satisfaction.

Visual Search and Augmented Reality: AI enables visual search capabilities, allowing customers to search for products using images rather than text. This technology uses computer vision algorithms to analyze images and identify similar products, providing customers with convenient and efficient product discovery. Additionally, augmented reality (AR) technology powered by AI allows customers to visualize products in real-world environments before making a purchase, enhancing the online shopping experience.

Supply Chain Optimization: AI can optimize inventory management, demand forecasting, and supply chain operations. By analyzing historical sales data, market trends, and external factors, AI algorithms can provide accurate demand forecasts, helping retailers optimize their inventory levels and reduce stockouts or excess inventory. AI can also optimize shipping and logistics operations, improving efficiency and reducing costs.

Fraud Detection and Security: AI algorithms can analyze patterns and anomalies in customer behavior to detect and prevent fraudulent activities, such as credit card fraud and identity theft. By continuously monitoring transactions and user data, AI-powered systems can identify suspicious activities and flag potential security risks, protecting both retailers and customers.

The integration of AI in retail enables retailers to offer personalized experiences, improve customer satisfaction, and optimize operations. However, it is essential to address privacy concerns, ensure transparent and ethical use of customer data, and provide robust cybersecurity measures to build trust with customers and maintain the integrity of AI-powered systems in the retail industry.

Manufacturing: Ai In Automation And Predictive Maintenance

Artificial intelligence (AI) is revolutionizing the manufacturing industry by driving automation and enabling predictive maintenance. AI-powered systems and technologies are transforming traditional manufacturing processes, enhancing productivity, efficiency, and overall equipment effectiveness. Here are some key applications of AI in manufacturing:

Automation and Robotics: AI enables intelligent automation by combining machine learning, computer vision, and robotics. AI-powered robots can perform repetitive and complex tasks with high precision and efficiency, reducing human errors and increasing production output. These robots can also adapt to changing production requirements, improving flexibility and agility on the factory floor.

Predictive Maintenance: AI-based predictive maintenance systems use advanced analytics and machine learning algorithms to monitor equipment conditions in real-time. By analyzing sensor data, performance metrics, and historical patterns, AI algorithms can identify potential equipment failures or malfunctions before they occur. This allows manufacturers to schedule maintenance activities proactively, minimizing downtime and optimizing equipment performance.

Quality Control and Inspection: AI can enhance quality control processes by automating inspection tasks. Machine vision systems equipped with AI algorithms can quickly and accurately identify defects, anomalies, or deviations from quality standards in manufactured products. This improves overall product quality and reduces the need for manual inspections, saving time and costs.

Supply Chain Optimization: AI can optimize supply chain operations by analyzing data from various sources, including demand forecasts, inventory levels, and supplier performance. AI algorithms can identify patterns, trends, and optimization opportunities, enabling manufacturers to make data-driven decisions to improve inventory management, reduce lead times, and enhance overall supply chain efficiency.

Energy Efficiency: AI can optimize energy consumption in manufacturing facilities by analyzing real-time data from sensors, meters, and other sources. AI algorithms can identify energy usage patterns, detect inefficiencies, and provide recommendations for energy-saving measures. By optimizing energy consumption, manufacturers can reduce costs, minimize their environmental footprint, and comply with sustainability goals.

Product Design and Innovation: AI technologies, such as generative design and simulation, can assist manufacturers in creating innovative and optimized product designs. AI algorithms can generate multiple design iterations based on specified parameters, enabling manufacturers to explore new possibilities and find the most efficient and cost-effective designs.

The integration of AI in manufacturing holds tremendous potential for improving productivity, quality, and operational efficiency. However, it is crucial for manufacturers to address challenges such as data security, workforce upskilling, and ensuring ethical and responsible use of AI technologies. By embracing AI, manufacturers can unlock new levels of competitiveness, agility, and innovation in the ever-evolving manufacturing landscape.

V. Ethical Considerations in Artificial Intelligence

Bias And Fairness In Ai Algorithms

Bias and fairness in AI algorithms are critical considerations in the development and deployment of artificial intelligence systems. Despite their potential benefits, AI algorithms can inadvertently perpetuate biases and inequalities present in the data they are trained on. Here are some key aspects related to bias and fairness in AI algorithms:

Data Bias: AI algorithms learn from historical data, which may contain biases reflecting societal inequalities and prejudices. If the training data is biased, the AI algorithm may make biased decisions or predictions. For example, in hiring or loan approval systems, biased training data can lead to discrimination against certain groups based on gender, race, or other protected characteristics.

Algorithmic Bias: Even with unbiased training data, the algorithms themselves may introduce biases. This can occur due to the choice of features, the model architecture, or the optimization process. Biased algorithms can result in unfair outcomes or reinforce existing disparities in areas such as criminal justice, healthcare, or financial services.

Fairness Metrics: Assessing fairness in AI algorithms requires defining appropriate fairness metrics. These metrics can capture various aspects of fairness, such as demographic parity (equal outcomes across different groups), equalized odds (equal false positive and false negative rates), or disparate impact (avoiding disproportionate adverse effects on protected groups). Selecting the right fairness metrics depends on the specific context and goals of the AI system.

Mitigating Bias: Addressing bias and promoting fairness in AI algorithms requires a combination of technical and ethical approaches. This includes carefully curating training data to minimize bias, auditing and testing algorithms for fairness, and incorporating fairness considerations throughout the entire development lifecycle. Regular monitoring and evaluation of AI systems in real-world settings are also crucial to identify and rectify any unintended biases or unfair outcomes.

Ethical and Responsible AI: Ensuring fairness goes beyond technical considerations. It involves making ethical and responsible choices throughout the AI development process. This includes promoting diversity and inclusion in AI teams, involving domain experts and impacted communities in decision-making, and transparently communicating the limitations and potential biases of AI systems to users and stakeholders.

Addressing bias and fairness in AI algorithms is an ongoing challenge that requires collaboration between data scientists, ethicists, policymakers, and society at large. It involves a multidisciplinary approach that combines technical expertise, ethical considerations, and a commitment to social responsibility. By striving for fairness and mitigating biases, we can harness the full potential of AI to create positive and equitable outcomes for individuals and communities.

Privacy And Security Concerns

Privacy and security concerns are significant considerations in the context of artificial intelligence (AI) systems. As AI technology advances and becomes more pervasive, it collects and processes large amounts of personal data, raising potential risks to individuals' privacy. Here are key aspects related to privacy and security concerns in AI:

Data Privacy: AI algorithms rely on extensive data, often including personal and sensitive information. It is essential to handle this data with utmost care to protect individuals' privacy rights. Data anonymization, encryption, and secure storage are crucial measures to safeguard personal information and prevent unauthorized access.

Informed Consent: Collecting and using personal data for AI purposes should be based on informed consent. Users should have a clear understanding of how their data will be used, who will have access to it, and what rights they have regarding their data. Transparent disclosure and user-friendly consent mechanisms are important to ensure individuals have control over their personal information.

Security Vulnerabilities: AI systems are not immune to security vulnerabilities, and malicious actors can exploit weaknesses to gain unauthorized access or manipulate the system. Robust security measures, such as secure coding practices, regular software updates, and system audits, are necessary to protect AI systems from cyber threats.

Algorithmic Transparency: The inner workings of AI algorithms can be complex and difficult to interpret. However, ensuring transparency in AI systems is crucial for understanding how decisions are made, detecting biases, and addressing potential discriminatory outcomes. Striving for algorithmic transparency can enhance accountability and trust in AI applications.

Ethical Use of AI: Privacy and security concerns extend beyond technical measures. Organizations and developers have a responsibility to use AI technology ethically and responsibly. This includes adhering to privacy regulations, establishing data governance frameworks, and conducting privacy impact assessments to identify and mitigate risks.

Regulatory Frameworks: Governments and regulatory bodies are increasingly recognizing the importance of privacy and security in AI. Implementing robust legal and regulatory frameworks can help ensure that AI applications comply with privacy laws, protect individuals' rights, and establish accountability for any misuse of personal data.

Balancing the potential benefits of AI with privacy and security concerns requires a comprehensive and proactive approach. It involves adopting privacy-by-design principles, fostering a culture of data protection, and engaging in ongoing dialogue between technology developers, policymakers, and society to address emerging challenges and protect individuals' privacy in the AI era.

Transparency And Explainability In Ai Decision-Making

Transparency and explainability in AI decision-making refer to the ability to understand and provide clear explanations for the decisions made by AI systems. As AI becomes more complex and sophisticated, there is a growing need to ensure that AI algorithms and models are transparent, accountable, and can be explained to both users and affected individuals. Here are key aspects related to transparency and explainability in AI decision-making:

Interpretable Models: AI models should be designed to be interpretable, meaning that the decision-making process can be understood and traced back to the input data and the underlying rules or patterns. Transparent algorithms, such as decision trees or linear models, are more easily interpretable compared to complex black-box models like deep neural networks.

Model Explanations: Providing explanations for AI decisions is crucial for building trust and understanding. Techniques such as feature importance analysis, rule extraction, or generating textual explanations can help explain how AI models arrived at specific outcomes. These explanations help users and stakeholders comprehend the reasoning behind the decisions made by AI systems.

Fairness and Bias Mitigation: Transparency and explainability play a vital role in identifying and addressing biases and fairness concerns in AI decision-making. By understanding how AI models make decisions, it becomes possible to detect and mitigate biases that may lead to discriminatory outcomes. Transparent AI systems allow for scrutiny and accountability, enabling corrective actions to be taken.

Auditing and Compliance: Transparent AI systems facilitate auditing and compliance processes. Organizations can track and review the decision-making process, ensuring that AI algorithms align with legal and ethical requirements. Auditing also helps identify potential issues or biases that may arise from the data used or the decision-making process itself.

User Trust and Acceptance: Transparency and explainability enhance user trust and acceptance of AI systems. When individuals can understand and follow the decision-making process, they are more likely to trust the outcomes. Transparent AI can also foster user engagement and allow individuals to make informed choices or challenge decisions when necessary.

Regulatory and Legal Considerations: Increasingly, regulatory frameworks are being established to require transparency and explainability in certain AI applications, particularly those with high stakes, such as healthcare or finance. Compliance with regulations and legal requirements necessitates transparency in AI decision-making to ensure accountability and protect individual rights.

Promoting transparency and explainability in AI decision-making is crucial for building trust, addressing bias, and ensuring ethical and responsible use of AI technology. It empowers users, promotes accountability, and helps mitigate potential risks and challenges associated with AI deployment in various domains.

Ensuring Responsible And Ethical Ai Practices

Ensuring responsible and ethical AI practices is essential to harness the benefits of artificial intelligence while mitigating potential risks and challenges. Here are key considerations for promoting responsible and ethical AI:

Fairness and Bias Mitigation: Strive to eliminate biases and ensure fairness in AI systems. Carefully examine the data used to train AI models to prevent bias from being propagated. Regularly evaluate and monitor AI systems for fairness across different demographic groups and make necessary adjustments to mitigate biases.

Privacy and Data Protection: Respect user privacy and handle personal data with care. Adhere to data protection regulations and implement robust security measures to safeguard sensitive information. Use anonymization and encryption techniques to protect user privacy while extracting meaningful insights.

Transparency and Explainability: Foster transparency and provide clear explanations for AI decisions. Make efforts to ensure that AI algorithms and models are interpretable, allowing users to understand the decision-making process and the factors that contribute to outcomes. Promote open dialogue and communication about AI systems' limitations and potential biases.

Accountability and Governance: Establish clear lines of accountability for AI systems. Define roles and responsibilities for monitoring, evaluating, and addressing any issues that may arise. Implement governance frameworks that promote responsible AI practices and provide mechanisms for reporting concerns or violations.

Human-Centered Design: Prioritize the needs and well-being of humans in AI development. Involve diverse stakeholders, including domain experts and end-users, in the design and evaluation process. Incorporate ethical considerations into the development cycle and ensure that AI systems align with societal values and norms.

Continuous Monitoring and Evaluation: Regularly assess the performance and impact of AI systems. Monitor their behavior and outcomes to identify any unintended consequences or biases. Implement mechanisms for ongoing evaluation and improvement, and be responsive to feedback from users and affected individuals.

Ethical Use and Decision-Making: Consider the broader ethical implications of AI applications and their potential impact on society. Ensure that AI systems are deployed in ways that align with ethical guidelines and respect human rights. Exercise caution in high-stakes domains, such as healthcare or criminal justice, where AI decisions can have significant consequences.

Collaboration and Knowledge Sharing: Foster collaboration among researchers, practitioners, policymakers, and other stakeholders to share knowledge and best practices. Engage in discussions and initiatives focused on responsible AI to collectively address challenges and promote ethical standards.

By embracing responsible and ethical AI practices, we can harness the transformative potential of artificial intelligence while upholding principles of fairness, transparency, accountability, and societal well-being. It requires a holistic approach involving technical expertise, ethical considerations, and collaboration across disciplines and sectors to navigate the evolving landscape of AI responsibly.

VI. Advancements and Future Trends
in Artificial Intelligence

Reinforcement Learning And Autonomous Systems

Reinforcement learning is a subfield of artificial intelligence that focuses on training autonomous systems to make decisions and take actions in dynamic environments. In reinforcement learning, an agent interacts with its environment, learning from trial and error to maximize a reward signal.

Autonomous systems that leverage reinforcement learning have the ability to learn and adapt to changing circumstances without explicit programming. These systems can be applied to a wide range of domains, including robotics, self-driving cars, gaming, and more. By using reinforcement learning algorithms, autonomous systems can learn complex behaviors and make intelligent decisions based on the feedback they receive from the environment.

One key advantage of reinforcement learning is its ability to handle situations where there is no pre-existing dataset or labeled examples. Instead, the agent learns by exploring the environment, receiving feedback in the form of rewards or penalties, and adjusting its actions accordingly. Through repeated interactions, the agent refines its decision-making policy and improves its performance over time.

However, there are challenges associated with reinforcement learning and autonomous systems. Training an agent through trial and error can be time-consuming and computationally expensive. It requires careful design of reward structures and exploration strategies to ensure effective learning. Balancing exploration and exploitation is crucial to avoid getting stuck in suboptimal solutions or local optima.

Another challenge is ensuring the safety and ethical behavior of autonomous systems. As these systems learn and make decisions independently, it is important to define appropriate constraints and regulations to prevent unintended consequences or harmful actions. Ensuring transparency and interpretability of the learned policies is crucial for building trust and understanding the decision-making process of autonomous systems.

Despite these challenges, reinforcement learning and autonomous systems hold great promise in enabling intelligent, adaptive, and autonomous behavior in various domains. Ongoing research and advancements in algorithms, hardware, and safety measures continue to drive progress in this field, opening up new possibilities for creating sophisticated and capable autonomous agents.

Generative Ai: Creativity And Artistry

Generative AI is a fascinating field within artificial intelligence that focuses on creating models and systems capable of generating new content, such as images, music, text, and even entire virtual worlds. This branch of AI combines creativity and artistry with advanced algorithms and deep learning techniques to produce original and unique outputs.

One of the key aspects of generative AI is its ability to learn from existing data and then generate new content that resembles the patterns and styles observed in the training data. This can be done through various approaches, including generative adversarial networks (GANs), variational autoencoders (VAEs), and recurrent neural networks (RNNs).

In the realm of art and creativity, generative AI has shown remarkable capabilities. It can generate realistic and visually stunning images, compose unique pieces of music, and even produce engaging and coherent pieces of writing. These AI-generated works can range from abstract and imaginative to highly realistic and detailed, depending on the training data and the algorithms used.

Generative AI is not only limited to replicating existing styles and patterns but also has the potential to explore new and innovative ideas. By combining different datasets or introducing random variations, generative AI models can produce novel and unexpected outputs, pushing the boundaries of creativity.

However, there are ongoing discussions and ethical considerations surrounding the use of generative AI in art. Questions about authorship, ownership, and the role of human creativity arise when AI systems are capable of producing original works. Nevertheless, generative AI offers exciting opportunities for artists, designers, and creatives to explore new avenues of expression and collaboration, blending human imagination with the power of AI.

As generative AI continues to advance, we can expect to witness even more astonishing and inspiring creations. From generating realistic landscapes to composing symphonies, generative AI is opening up new possibilities for human-artificial intelligence collaboration and pushing the boundaries of what is creatively possible.

Ai In Robotics And Human-Robot Interaction

AI in robotics and human-robot interaction represents a groundbreaking field that aims to integrate artificial intelligence into robotic systems, enabling them to interact and collaborate with humans in various settings. This convergence of AI and robotics has the potential to revolutionize industries, enhance productivity, and transform the way we live and work.

One of the key applications of AI in robotics is in autonomous robots. These robots are equipped with AI algorithms that enable them to perceive and understand their environment, make intelligent decisions, and perform complex tasks without constant human intervention. They can navigate through dynamic environments, interact with objects, and adapt their behaviors based on changing circumstances.

AI also plays a crucial role in human-robot interaction (HRI), which focuses on designing robots that can effectively communicate, understand, and respond to human commands, gestures, and emotions. Natural language processing and computer vision techniques are used to enable robots to understand human speech, interpret facial expressions, and engage in meaningful conversations. This facilitates seamless collaboration and cooperation between humans and robots in various domains, such as healthcare, customer service, and manufacturing.

Moreover, AI enables robots to learn from human demonstrations and experiences, allowing them to acquire new skills and adapt to different situations. Reinforcement learning techniques, for example, enable robots to improve their performance through trial and error and optimize their actions based on feedback from the environment.

However, as AI-powered robots become more prevalent, ethical considerations come into play. Ensuring the safety, privacy, and ethical use of robots in human-robot interactions is of utmost importance. Regulations and guidelines need to be established to address concerns such as data privacy, security, and the impact of automation on the workforce.

Overall, AI in robotics and human-robot interaction holds immense potential to enhance our daily lives, revolutionize industries, and contribute to advancements in science and technology. By combining the power of artificial intelligence with physical robotic systems, we can create intelligent machines that augment human capabilities, enable new forms of collaboration, and pave the way for a future where robots and humans work together in harmony.

Ai And The Internet Of Things (Iot)

AI and the Internet of Things (IoT) are two transformative technologies that are increasingly interconnected, driving innovation and shaping the future of various industries. The combination of AI and IoT brings together intelligent data processing and connectivity, enabling the creation of smart and autonomous systems that can revolutionize our homes, cities, and workplaces.

The IoT refers to the vast network of interconnected physical devices, sensors, and actuators embedded with software and network connectivity. These devices collect and exchange data, creating a wealth of information about our environment and activities. AI complements the IoT by providing the means to analyze and make sense of this vast amount of data, extracting valuable insights and enabling intelligent decision-making.

AI algorithms can process and interpret IoT data, identifying patterns, trends, and anomalies that would be difficult for humans to detect. This allows for real-time monitoring, predictive analytics, and proactive decision-making in various domains. For example, in smart homes, AI-powered IoT systems can learn users' preferences and adjust lighting, temperature, and other settings accordingly. In smart cities, IoT sensors can collect data on traffic patterns, air quality, and energy consumption, while AI algorithms can optimize resource allocation and improve urban planning.

Moreover, AI can enhance IoT devices and systems by enabling them to learn and adapt. Machine learning algorithms can be applied to IoT data to improve the performance and efficiency of devices, as well as to enable predictive maintenance. AI-powered IoT systems can automatically adjust their behavior based on changing conditions and user preferences, leading to more personalized and efficient experiences.

However, the integration of AI and IoT also brings challenges and considerations. Security and privacy become critical, as IoT devices collect and transmit sensitive data. Robust cybersecurity measures must be in place to protect against data breaches and unauthorized access. Additionally, ethical considerations arise when it comes to the use of AI in IoT, such as ensuring transparency and accountability in decision-making processes.

In conclusion, the combination of AI and IoT has the potential to transform our lives and reshape industries. By leveraging the power of AI to analyze and interpret IoT data, we can unlock new insights, optimize processes, and create intelligent systems that enhance efficiency, convenience, and sustainability. However, it is crucial to address security, privacy, and ethical concerns to ensure the responsible and beneficial integration of AI and IoT technologies.

Quantum Computing And Its Impact On Ai

Quantum computing is an emerging field of technology that has the potential to revolutionize various areas, including artificial intelligence (AI). Unlike classical computers, which use bits to represent and process information as either 0 or 1, quantum computers utilize quantum bits or qubits, which can exist in multiple states simultaneously due to the principles of quantum mechanics.

The impact of quantum computing on AI lies in its ability to perform complex calculations and solve problems at a significantly faster rate than classical computers. This has the potential to enhance the capabilities of AI algorithms and models, enabling more sophisticated data analysis, pattern recognition, and optimization.

One area where quantum computing can have a significant impact on AI is in the realm of optimization problems. Many real-world problems involve finding the best possible solution among countless options, such as route optimization, portfolio management, or resource allocation. Quantum algorithms, such as the quantum approximate optimization algorithm (QAOA) or the quantum annealing algorithm, offer the potential to outperform classical optimization algorithms, leading to more efficient and effective solutions.

Quantum computing can also enhance machine learning algorithms by enabling the exploration of large, high-dimensional datasets and feature spaces. Quantum machine learning algorithms, such as quantum support vector machines (QSVM) or quantum neural networks, leverage the unique properties of quantum computing to process and analyze data in ways that classical computers cannot. This opens up new possibilities for tackling complex machine learning tasks and improving the accuracy and efficiency of AI models.

However, it's important to note that quantum computing is still in its early stages of development, and practical, scalable quantum computers that can outperform classical computers in AI tasks are not yet readily available. Overcoming the challenges of quantum error correction, decoherence, and scaling up the number of qubits is crucial for realizing the full potential of quantum computing in AI.

In conclusion, quantum computing holds promise for advancing the field of AI by providing faster computational power and enabling novel algorithms and approaches. As quantum technologies continue to mature, they have the potential to drive breakthroughs in various AI applications, leading to more powerful and efficient AI systems. Nonetheless, further research, development, and refinement are necessary to fully integrate quantum computing into the realm of AI and unlock its transformative potential.

VII. Impact of Artificial Intelligence on Society

Job Displacement And Workforce Transformation

The advancement of artificial intelligence (AI) and automation technologies has raised concerns about job displacement and the transformation of the workforce. As AI systems become more sophisticated and capable of performing tasks traditionally done by humans, there is a potential for certain jobs to be replaced by machines or algorithms.

Job displacement occurs when tasks that were previously performed by humans are taken over by AI systems. This can affect a wide range of industries and job roles, from manufacturing and customer service to data analysis and even professional services. Jobs that involve routine, repetitive tasks are particularly vulnerable to automation.

However, it's important to note that while some jobs may be replaced, new jobs and opportunities can also emerge as a result of AI and automation. As technology advances, new industries and job roles are created, requiring skills that complement and work alongside AI systems. For example, the development, maintenance, and management of AI systems require specialized skills in areas such as data science, machine learning, and AI ethics.

Workforce transformation is the process of adapting to the changing job landscape and acquiring new skills to remain relevant in the age of AI. It involves upskilling and reskilling workers to equip them with the knowledge and abilities needed to work alongside AI systems or transition to new roles that leverage their unique human capabilities. This could involve developing skills in areas such as critical thinking, creativity, problem-solving, and emotional intelligence, which are less likely to be automated.

To address job displacement and support workforce transformation, collaboration is crucial between governments, educational institutions, employers, and individuals. Governments can play a role in creating policies and programs that facilitate the reskilling and upskilling of workers. Educational institutions can adapt their curricula to include AI-related skills and promote lifelong learning. Employers can invest in training programs and provide opportunities for employees to acquire new skills. Individuals can take initiative in upgrading their skills and staying adaptable to the changing job landscape.

Overall, while job displacement is a valid concern, it is important to view AI and automation as tools that can augment human capabilities rather than completely replace them. By embracing the opportunities presented by AI, investing in skills development, and fostering a culture of continuous learning, we can navigate the workforce transformation and create a future where humans and AI systems work together synergistically.

Education And Ai: Preparing For The Future

Education plays a crucial role in preparing individuals for the future, particularly in the context of AI. As AI continues to advance and reshape various industries, it is essential that education systems adapt to equip students with the skills and knowledge needed to thrive in an AI-driven world.

One key aspect of education in the AI era is fostering a strong foundation in STEM subjects (science, technology, engineering, and mathematics). These disciplines provide the fundamental knowledge and analytical thinking skills necessary to understand and work with AI technologies. Students should be encouraged to explore areas such as computer science, data analysis, and robotics, which form the building blocks of AI.

Additionally, education should focus on cultivating essential human skills that complement AI capabilities. These include critical thinking, creativity, problem-solving, emotional intelligence, and ethical reasoning. These skills are uniquely human and can enhance collaboration, innovation, and adaptability in the face of AI-driven changes.

Integration of AI into educational practices can also provide valuable learning experiences. AI-powered tools and platforms can support personalized learning, adaptive assessment, and intelligent tutoring systems, tailoring education to individual student needs. This can foster engagement, improve learning outcomes, and prepare students for a future where AI technologies are prevalent.

Furthermore, education must address the ethical implications of AI. Teaching students about AI ethics, privacy, and fairness helps them understand the societal impact of AI and empowers them to make informed decisions when developing or interacting with AI systems.

Collaboration between educational institutions, industry

professionals, and policymakers is essential in shaping educational curricula and programs. Partnerships can facilitate the exchange of knowledge, resources, and best practices, ensuring that education remains relevant and responsive to the evolving AI landscape.

Ultimately, education in the AI era should focus on equipping individuals with a combination of technical expertise, critical thinking skills, and ethical awareness. By fostering a strong educational foundation and a lifelong learning mindset, individuals can navigate the opportunities and challenges presented by AI, contribute meaningfully to society, and shape a future that leverages AI for the benefit of all.

Ai And Social Implications: Ethics, Policy, And Governance

The advancement of artificial intelligence (AI) brings with it a range of social implications that need to be carefully considered. Ethics, policy, and governance play crucial roles in ensuring that AI technologies are developed, deployed, and used in a responsible and beneficial manner.

Ethics in AI involves addressing ethical considerations and ensuring that AI systems are designed and used in ways that align with moral principles and societal values. This includes issues such as fairness, accountability, transparency, privacy, and the potential for bias. Ethical frameworks and guidelines can help guide the development and use of AI, ensuring that it promotes human well-being and avoids harm.

Policy and governance frameworks are necessary to regulate and manage the use of AI technologies. Policymakers need to establish clear guidelines and regulations to address various aspects of AI, including data protection, cybersecurity, liability, and intellectual property rights. These policies should balance innovation and societal benefits with the need to mitigate risks and protect individuals' rights.

Collaboration between governments, industry, academia, and civil society is crucial in shaping effective policies and governance mechanisms. Open and inclusive discussions, involving multiple stakeholders, can help address the complexities of AI and ensure that policies are comprehensive, balanced, and reflective of diverse perspectives.

International cooperation is also important in addressing the global impact of AI. Harmonizing regulations, sharing best practices, and establishing common ethical standards can foster trust, enable responsible AI development, and address challenges that cross national boundaries.

Moreover, public awareness and education about AI and its implications are essential. This includes initiatives to promote digital literacy, raise awareness about AI's potential benefits and risks, and foster public dialogue on ethical considerations and policy decisions.

By addressing the social implications of AI through ethical considerations, robust policies, and effective governance, we can harness the transformative power of AI while minimizing potential risks. This requires a proactive and collaborative approach that ensures AI technologies are developed and deployed in ways that align with our shared values and contribute to a more inclusive, equitable, and sustainable future.

Ai For Social Good: Addressing Global Challenges

AI has the potential to address various global challenges and contribute to social good in numerous ways. By leveraging the power of AI technologies, we can tackle complex problems, improve decision-making processes, and create innovative solutions to benefit individuals, communities, and society as a whole.

In healthcare, AI can revolutionize disease diagnosis and treatment. AI algorithms can analyze vast amounts of medical data, assisting doctors in making accurate diagnoses and recommending personalized treatment plans. It can also contribute to drug discovery, accelerate medical research, and improve patient outcomes.

In education, AI can enhance learning experiences by providing personalized and adaptive learning platforms. Intelligent tutoring systems can assess students' strengths and weaknesses, tailor instructional materials, and provide targeted feedback. AI-powered educational tools can also assist teachers in developing effective teaching strategies and optimizing educational resources.

AI can play a significant role in addressing environmental challenges. For example, it can help optimize energy consumption, predict and mitigate the impact of natural disasters, and support conservation efforts. AI algorithms can analyze data from sensors and satellite imagery to monitor environmental changes and assist in sustainable resource management.

In poverty alleviation and economic development, AI can promote financial inclusion, provide access to essential services in underserved areas, and support entrepreneurship. AI-powered chatbots and mobile applications can enable access to information, financial services, and healthcare in remote regions, helping bridge the digital divide.

AI also has the potential to enhance public safety and security. Facial recognition and video analytics can aid in crime prevention, disaster response, and cybersecurity. AI-powered surveillance systems can analyze patterns, detect anomalies, and alert authorities to potential threats.

To ensure the responsible use of AI for social good, collaboration between AI developers, policymakers, researchers, and communities is essential. Ethical considerations, transparency, and inclusiveness should guide the development and deployment of AI technologies to address societal challenges effectively. It is crucial to mitigate biases, promote fairness, and ensure that AI solutions are accessible and beneficial to all.

By harnessing AI's capabilities and directing them towards addressing global challenges, we can make significant strides in building a more sustainable, equitable, and prosperous future for all.

VIII. Conclusion

Recap Of Key Insights And Takeaways

Throughout the book, we have explored the transformative power of artificial intelligence (AI) and its impact on various aspects of our lives. Here are some key insights and takeaways:

AI is a broad field encompassing different technologies and approaches that aim to mimic human intelligence and enhance decision-making processes.

Machine learning, deep learning, and natural language processing are key branches of AI that enable systems to learn from data, process complex information, and understand and generate human language.

AI has significant applications across industries such as healthcare, finance, transportation, retail, manufacturing, and more. It has the potential to improve diagnostics, treatment, customer experiences, automation, and predictive maintenance, among other areas.

Ethical considerations, bias mitigation, privacy, and transparency are crucial aspects of AI development and deployment. It is important to ensure responsible and ethical AI practices to avoid unintended consequences and promote fairness.

AI's impact extends beyond technology to social and economic realms. It can lead to job displacement and workforce transformation, requiring us to adapt and reskill to remain relevant in the AI-driven world.

Education plays a crucial role in preparing individuals for the AI future. Developing AI literacy and promoting interdisciplinary education can equip us with the skills and knowledge needed to leverage AI technologies effectively.

AI has implications for policy, governance, and ethics. It is important to establish frameworks and regulations that address societal concerns, promote accountability, and ensure the responsible and beneficial use of AI.

AI has immense potential for social good, addressing global challenges, and creating positive change. It can contribute to healthcare advancements, environmental sustainability, poverty alleviation, public safety, and more.

Collaboration, inclusivity, and multidisciplinary approaches are key to harnessing the benefits of AI and addressing its challenges. Engaging stakeholders from various fields and fostering a dialogue between academia, industry, policymakers, and communities is vital.

As AI continues to evolve, it is essential to stay informed, keep up with advancements, and participate in shaping the future of AI. By embracing AI's transformative potential and adopting responsible practices, we can create a more inclusive, equitable, and prosperous society.

Remember, AI is a powerful tool that requires careful stewardship to ensure its positive impact on humanity. By understanding its capabilities, limitations, and ethical considerations, we can leverage AI's potential for the betterment of individuals, communities, and the world at large.

Encouragement For Further Exploration Of Ai

As you conclude your journey through the world of artificial intelligence, I encourage you to continue exploring and engaging with this fascinating field. Here are some ways to further your exploration of AI:

Stay updated: AI is a rapidly evolving field with new advancements and research emerging constantly. Stay connected with reputable sources, such as research papers, conferences, and industry publications, to keep abreast of the latest developments.

Join communities: Engage with AI communities, both online and offline, to connect with like-minded individuals, researchers, practitioners, and enthusiasts. Participate in forums, discussion groups, and social media platforms dedicated to AI, where you can exchange ideas, ask questions, and learn from others.

Learn and upskill: Consider taking courses, attending workshops, or pursuing certifications in AI-related topics. Many educational institutions and online platforms offer AI courses catering to different skill levels and interests. Deepening your knowledge and acquiring practical skills will enable you to apply AI concepts effectively in various domains.

Collaborate and network: Seek opportunities to collaborate with professionals working in AI or related fields. Engage in projects, attend AI meetups and conferences, and build connections with individuals who share your passion for AI. Collaborative efforts can lead to valuable insights, learning experiences, and even potential career opportunities.

Experiment and innovate: Don't be afraid to dive into hands-on AI projects. Apply your knowledge to real-world problems, develop AI applications, or contribute to open-source AI projects. Embrace a mindset of experimentation and innovation to deepen your understanding and explore the creative possibilities of AI.

Engage in ethical discussions: AI raises important ethical questions and considerations. Join conversations and debates on the ethical implications of AI, and contribute your insights to foster responsible and inclusive AI practices. Engaging in ethical discussions will help shape the future of AI in a way that aligns with our values and societal needs.

Contribute to AI for social good: Explore ways to leverage AI for social impact. Identify areas where AI can address pressing global challenges, such as healthcare, education, climate change, or social justice. Collaborate with organizations or start your own initiatives that use AI for positive change, making a difference in the world.

Remember, AI is a tool that holds immense potential, and it is our collective responsibility to ensure its ethical and responsible use. By continuing to explore, learn, collaborate, and contribute to the field of AI, you can be part of shaping its future and unlocking its transformative possibilities. Embrace the journey with curiosity, open-mindedness, and a commitment to making AI a force for good.

Embracing The Intelligent Future

As you embrace the intelligent future, you are poised to witness the remarkable advancements and opportunities that artificial intelligence brings. Here are some ways to embrace and make the most of the intelligent future:

Embrace lifelong learning: AI is a dynamic and rapidly evolving field. Stay curious and committed to continuous learning. Explore new AI technologies, algorithms, and applications. Expand your knowledge and skills to adapt to the changing landscape and stay relevant in a world driven by intelligent systems.

Embrace collaboration: AI thrives in an ecosystem of collaboration. Seek opportunities to collaborate with professionals from diverse backgrounds. By bringing together different perspectives, skills, and expertise, you can unlock new insights, foster innovation, and tackle complex challenges more effectively.

Embrace interdisciplinary approaches: AI intersects with various disciplines, such as computer science, mathematics, psychology, ethics, and more. Embrace interdisciplinary approaches to gain a holistic understanding of AI and its impact on society. Foster connections between AI and other fields to unlock synergies and address multifaceted problems.

Embrace ethical considerations: AI raises important ethical questions. Be mindful of the potential biases, privacy concerns, and social implications associated with AI. Champion ethical practices, transparency, and accountability in the development and deployment of AI systems. Strive for fairness, inclusivity, and societal well-being in all AI-related endeavors.

Embrace human-centric design: As AI becomes more integrated into our lives, prioritize human needs and experiences. Design AI systems that are intuitive, user-friendly, and enhance human capabilities. Consider the social, cultural, and ethical aspects of AI to ensure that technology serves humanity's best interests.

Embrace responsible AI leadership: As AI continues to shape industries and societies, embrace the role of a responsible AI leader. Advocate for policies that promote the responsible and ethical use of AI. Engage in conversations, contribute your expertise, and influence the development of AI governance frameworks that ensure its benefits are accessible to all.

Embrace the positive potential of AI: While acknowledging the challenges, focus on the positive potential of AI. Explore its applications in addressing global issues such as healthcare, climate change, poverty, and education. Embrace AI as a tool to augment human capabilities, enhance productivity, and create a more sustainable and inclusive future.

By embracing the intelligent future with an open mind, a commitment to ethical practices, and a focus on human well-being, you can shape the trajectory of AI and harness its transformative power for the benefit of individuals, communities, and the world at large. Embrace the opportunities that lie ahead, and let the intelligent future unfold with optimism and purpose.